No Sweat Addition

Knowing and Understanding Addition

No Sweat Addition

Knowing and Understanding Addition

by

Carol Pirog

Anchor Book Press · Palatine

No Sweat Addition: Knowing and Understanding Addition

Anchor Book Press
440 W Colfax Street, Unit 1132, Palatine, IL 60078
ISBN: 978-1949109542

Printed in the United States

Introduction

I firmly believe that the biggest problem for students struggling with math is a lack of number sense and understanding of the prerequisites for the math they are currently plagued with. With all the pressure on teachers to complete the curriculum, the class often moves on before students have a strong foundation of basic math concepts. After all, if a student lacks a solid understanding of the base ten system, he or she is going to have difficulty reading numbers, adding and subtracting with regrouping, decimals, etc. If a student does not have a basic understanding of fractions, understanding decimals is really going to be difficult. In today's fast-paced curriculum, there is little time to reteach concepts students do not grasp. But, regardless of who is to blame, it is the students who suffer; it is students who think they will never understand math.

Another big issue for students is memorization. Many times, we ask students to memorize something without understanding, in which case they lack the skills to recognize errors. Other times, we avoid having students memorize things because we want them to understand. Math is not either/or. Students need both – to understand and memorize. Research has shown that students who are most proficient in math memorized math facts and learned math concepts at the same time. (Yes, I have heard all the outlier stories of math gurus who never learned their math facts, but they are just that – outliers, not what you can expect for most children.) Memorizing facts helps students develop math sense. They realize that 10 objects are 10 objects whether they have 3 objects added to 7 objects or they have 5 groups of 2 objects each.

Memorization is also necessary for math fluency. If a student spends so much time figuring out facts, it distracts from the main problem and makes it more difficult to understand. On the other hand, if a student memorizes facts and understands the operation, the information is helpful, allowing student to generalize in other situations and to solve complex problems without tediously calculating basic facts.

No Sweat Addition gives you the tools to help your child master basic addition facts and to understand the concept of addition.

Note: If your child is in 4th grade or higher, put the addition facts aside. Do multiplication first, as it is more important for older children. They can figure out addition facts much fast than multiplication facts. Once the child had mastered multiplication, then go back to addition.

No one is perfect - that’s why pencils have erasers.

Wolfgang Riebe

What Do the Standards Actually Say?

There has always been a lot of disagreement over whether students need to learn facts or not. For those of you struggle with Common Core, this is not a new issue. However, with Common Core math, one of the ideas was that students would understand what they were doing rather than just rote memorization of math facts without understanding. Having said that, unbeknownst to many teachers, parents, and students, Common Core does not say students no longer need to memorize the facts. Let's look at what the first-grade standards actually say about memorizing addition and subtraction facts. We are going to use a Common Core standard, but the standards are similar in states without Common Core. What people do not realize is Common Core did not change what students needed to learn. The big uproar is caused by other things that have been called Common Core, but that is a subject for another book. If it is something that interests you, check out my book, *Surviving Common Core Math.* When you read this standard, the thing to realize is that the difference in Common Core and non-Common Core standards is that it gives more information. In non-Common Core states, the standard usually consists of just the first sentence.

CCSS.MATH.CONTENT.1.OA.C.6
Add and subtract within 20, demonstrating fluency for addition and subtraction within 10. Use strategies such as counting on; making ten (e.g., 8 + 6 = 8 + 2 + 4 = 10 + 4 = 14); decomposing a number leading to a ten (e.g., 13 - 4 = 13 - 3 - 1 = 10 - 1 = 9); using the relationship between addition and subtraction (e.g., knowing that 8 + 4 = 12, one knows 12 - 8 = 4); and creating equivalent but easier or known sums (e.g., adding 6 + 7 by creating the known equivalent 6 + 6 + 1 = 12 + 1 = 13). (*You can find this and other State Standards on state board of education websites.)*

If we look at the standard, you will notice more than one goal. This is a Common Core goal, but states without Common Core have the same goals, but usually not written as one goal. In this standard, students should be able to add and subtract within 20, demonstrating fluency. Next is a list of strategies students can use to figure out answers they do not know that demonstrate an understanding of addition and subtraction.

When standards say **fluency** or **fluently**, this indicates students should know facts without counting on fingers or using other methods to figure out the answer. The students know the answer because they have memorized it. Goals in non-Common Core states will often just state in plain language that students should memorize facts. (The standards do not indicate how fast students can spit back the answer as all students process information at various speeds based on intelligence, emotional state, physical condition, etc.) Why is memorization needed? Most of us are familiar with reading fluency – the ability to quickly read a passage without needing to sound out most words. Fluency in reading allows a student to use available brain resources for understanding the meaning of the passage (reading comprehension) instead of decoding words. This does not mean the student has not learned strategies to decode a couple unknown words in a passage. The use of **fluency** in this 1st grade Common Core Math Standard shows us that **fluency** in math, while not as well known, is also expected. This is true in non-Common Core states, as well. The standard also includes ways to calculate unknown sums, based on understanding, in the event a basic fact is forgotten or was not learn. Most facts must be known, or the student will not be fluent. Math **fluency** plays the same role in math as reading fluency plays in reading. **Fluency**, in math and reading, improve comprehension. Students working difficult math problems are distracted from the problem, if they have to take time to use a calculator or figure out many basic facts, while students who are **fluent** in math facts can focus on problem solving.

What then is the difference between the rote memorization of facts 50 years ago and the memorization of facts today? Nothing, except Common Core memorization is not called 'rote memorization'. Rote is a descriptor added to give memorization a negative connotation. While 'rote memorization' has a specific meaning that implies memorization without understanding, today it is used indiscriminately for any memorization without regard to details that make a difference. Therefore, the goal of Common Core math is largely misunderstood. It is not to get rid of memorization, but to get rid of **rote** memorization with no understanding. As far as facts go, they were never meant to be taught without also teaching understanding, even fifty years ago. Look for old math books and you will see the pages with pictures clearly meant to help students understand the concepts of addition and subtraction. Common Core and other state standards want students to memorize facts and understand addition and subtraction, as was the goal fifty years ago.

How do we know this? It is a simple matter to check the standards. The first sentence of the previous standard says kids need to add and subtract fluently. In other words, memorize facts. The sentences that follow in the standard are about understanding addition and subtraction. States without Common Core do not have a problem with plainly stating students should memorize facts.

The need for understanding becomes evident during work with word problems when the students do not know whether to add or subtract. If students understood the concept, a word problem would clarify what needed to be done, not confuse students. Ask any teacher, students struggle the most with word problems. Therefore, Common Core emphasizes understanding, not because memorizing is not needed but students are struggling with understanding the most.

If you think I am being dramatic, consider this scenario. Little Johnnie is in 2nd grade. In second grade students learn to add larger numbers with regrouping. Little Johnnie knows some addition facts like 2 + 2 and 2 + 5, but he just can't remember many others. No problem, his teacher tells him. Use your number line or your hundreds chart. The teacher passes out a worksheet with 5 double-digit addition problems. Most of the class finishes quickly, but it takes little Johnnie a long time. As a matter of fact, he is the last one to finish because he had to use his number line to count for each column. As this happens time and again, the other students take notice and begin teasing Johnnie. Before long, Johnnie starts to hate math. He plays with his pencil and doodles on his paper. He starts bothering the girl next to him and refusing to do his math because he would rather peers think he is bad than for them to think he is dumb. All this could be avoided if someone just realized how important it was for Little Johnnie to learn his facts.

The strategies in this book will help children learn facts, easily. Even if they have difficulty with memorization.

How to Use this Book

Key Concepts for Parents: (Teachers see end pages for use in the classroom.)

- Fun, fun, fun – do all you can to keep this fun - use rewards, competitions, charts. Why? Children will be more engaged if they are having fun. Being engaged is necessary for any program to work.
- Pass one lesson before moving to the next lesson – high five, sticker, smiley face – acknowledge accomplishment – whatever works for your child. For the struggling child, celebrate the growth (more correct answers on the page than yesterday or last week).
- Teach new facts before you child completes the student practice pages. This is the time to figure out answers – have your child use various strategies to calculate the answers. See Chart at end of book for new facts or look at next lesson.
- Child should know facts from memory when they complete the student practice pages (Lessons A – Z and Mastery 1 – 5)
- If student is on any level for more than 2 or 3 days, practice new facts in isolation (Note: it is more effective to practice for 2 minutes 5 times a day than it is to do 1 ten-minute practice session)
- Students can complete in writing or orally (note: writing triggers memory areas of brain)
- Teach vocabulary – another major problem for students is not understanding the vocabulary
- One-size does not fit all, use additional strategies if your child is still struggling. (See back of book for additional strategies.)

To begin, copy page for child to use. If your child does not pass, copy the same lesson for the next day.

The top section of student practice page (Lessons A – Z) is understanding. Fold paper so that student only sees the bottom. The bottom section is memorization and should be timed. Timing is not critical, but it is used to keep students focused on the assignment. Depending on the age and manual dexterity of your child, anywhere from 30 – 60 seconds should be more than enough. Remember, it is not that the student can spit the answers back at lightning speed, but that the student is giving the answer from memory, not calculating the answer. Students can also answer orally. Although, research has shown that the physical act of writing triggers the memory area of the brain. Unless your child has a physical handicap, which makes writing difficult, writing will make the memorization process faster.

Key Ideas (teach these concepts as you work through the book) **for Student:**

- 3 + 5 is read as 3 plus 5
- 3 + 5 means 3 objects and 5 more objects
- Addition is reciprocal: 3 + 5 is the same answer as 5 + 3
- This is also the time to talk about adding things that are alike. You can add apples, or you can add oranges. If you need to add apples and oranges, you have to change the category to fruit. Talking about this now, paves the way for future math concepts like adding fractions with unlike denominators.

Let's Get Started

Before starting student practice pages work with child on understanding addition:

Day 1 (Understanding addition is finding the total number in 2 or more groups)

1. Work with manipulatives to help child understand – use counters, pennies, M & Ms, etc
2. Start with 3 + 5. Put 3 objects a group and 5 objects in a different group. (Make groups by using bowls, saucers, circles drawn on a sheet of paper). Put 3 objects in one group and have child count the objects (if your child cannot count to 20, he or she is not ready for this book. (Work on counting before continuing.) Write on paper or white board, "3 objects." Then make another group with 5. Write "5 objects." Have your child count all objects. Say, "The math equation to go with this problem is 3 + 5 = 8." Write 3 + 5 = 8. (If your child does not know what an equation is, now is the time to teach this information – a number sentence using numerals and symbols.)
3. Play around with some other problems: **3 + 4** or **2 + 6** or **3 + 6** (keep the numbers small, so your child is not struggling with counting and other concepts). When teaching understanding, it is best to use small, easily manipulated facts. This is so your child understands what is happening when he or she adds. **Understanding** is the difference between rote memorization and the goal of fluency which requires memorization of addition facts where your child understands what addition is.

Day 2 (Vocabulary & Signs)

1. Plus, equals, +, =
2. Write addition equations. Read as you point to each number and symbol, then have your child read. (Tell child what symbol is, if needed.) If the child is unable to read numbers, work on reading numbers before continuing.
3. Talk about what the words and signs mean. See the back of book for kid friendly definitions.

Day 3 (Identity Property of Addition)

1. First talk about identity. Identity for this activity is your name. My name is Carol. That is my identity. Repeat using your child's name. Ask the identity of various people your child knows.

2. Once your child seems to understand identity of people, use letters (that your child knows). This is the letter A. Its identity is A because that is what we call the letter. Move to numbers once you are sure your child understands the identity of letters.
3. Explain: The identity property (rule) of addition is when 0 is added to a number, you get the same number – the number of objects and the identity/name does not change.
4. Work with numbers plus 0.
5. 3 + 0 (Explain: You have 3 objects and you do not get anymore.) The math sentence for that statement is 3 + 0 = 3. You have 3 and you get zero (same as nothing) so you still have 3.
6. Try with 2 + 0.
7. Work with 2-digit numbers like 23 + 0 = 23. Compliment your child on the ability to add large numbers and zero. Confidence goes a long way when your child needs to work hard to understand. Competency in this easy concept is the place to start building that confidence.

Day 4 – Doubles

1. This is something fun to do, but not absolutely necessary before you start the student practice pages.
2. Over the next couple of days, work on remembering 1 or 2 doubles facts (1 + 1; 2 + 2; etc). Just do one or two a day so your child is not overwhelmed.

Day 5 – Start Lessons A to Z

1. Have your child complete the top. Go over it with the child to make sure it is correct (I find it easiest just to watch as child completes, but some children feel pressured when being watched). Reteach anything that is not correct.
2. Time child while doing the bottom section (remember – it is not about speed, but just to keep child focused and writing answers from memory).
3. Have child fix any incorrect answers. If child misses more than 2 problems, redo the same page tomorrow or until child can complete with no more than 2 wrong answers. Note: an answer is not wrong if the formation of letter is poor. If that is a problem, work on handwriting at another time.
4. If child needs to do a lesson more than 3 days, work on the problems that are wrong in isolation. You can do that through flashcards or having the child write the missed problem 3 time while saying it out loud (2 plus 5 equals 7). This introduces other modalities of learning.

Vocabulary Development:

1. Words: sum, total, plus, equals, addends
2. If your child struggles with understanding use a visual for practice

a. Each day, write an addition fact with the answer (vertically one day, horizontally the next)
b. Start out with just two words – Tell your child to circle the total, put a box around the plus sign (or check if child has difficulty drawing a box). Give directions one at a time.
c. Add two more words when your child knows total and plus sign – Tell your child to circle the total, put box around plus sign, draw arrow to equals sign.
d. Finally, put X on the addends (all words at the end)
e. After your child can identify all parts of the problem work on naming – point to something and ask what it is called.
f. Note: using sum and total will prepare child for word problems.

We are ready to begin with Lesson A. Remember, make it fun.

Note: Spelling is important. This is math, focus on details is essential. Point out that the child can copy the spelling from the example. Provide a resource sheet/card with numbers one to twenty and the words plus, equals, and objects.

In every job that must be done, there is an element of fun. You find the fun and ‘snap’ the job’s a game.

Mary Poppins

Lesson A

Level 1 **A**

Draw a picture:

Example: 5 + 0

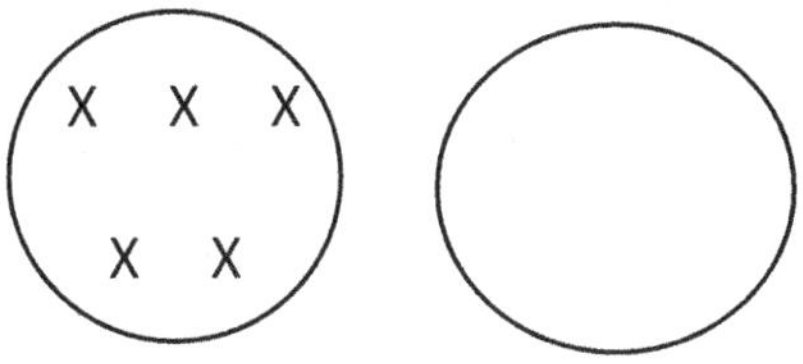

2 + 0

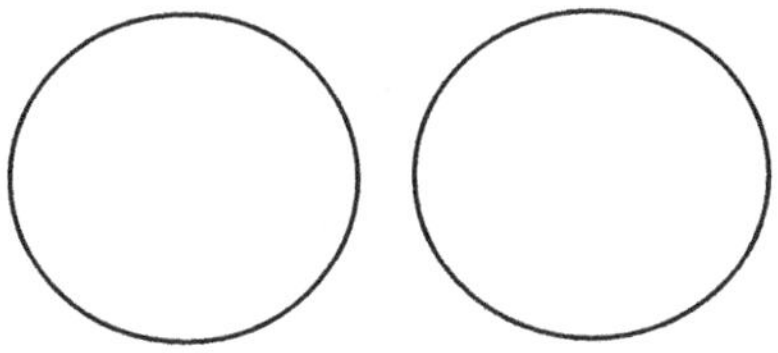

Write a word sentence (if writing is difficult, have child dictate word sentence to you):

Example: 3 + 0 <u>three objects plus zero objects</u>

4 + 0 ______________________________

Identity Property of Addition: Any number plus 0, equals the same number. **A**

2 + 0 = _____	5 + 0 = _____	9 + 0 = _____	0 + 1 = _____	6 + 0 = _____
0 + 7 = _____	3 + 0 = _____	0 + 0 = _____	0 + 8 = _____	0 + 10 = _____
7 + 0 = _____	8 + 0 = _____	0 + 3 = _____	0 + 4 = _____	0 + 6 = _____
0 + 5 = _____	0 + 2 = _____	4 + 0 = _____	1 + 0 = _____	10 + 0 = _____

Lesson B

B

Level 1

Draw a picture:

Example: 5 + 1

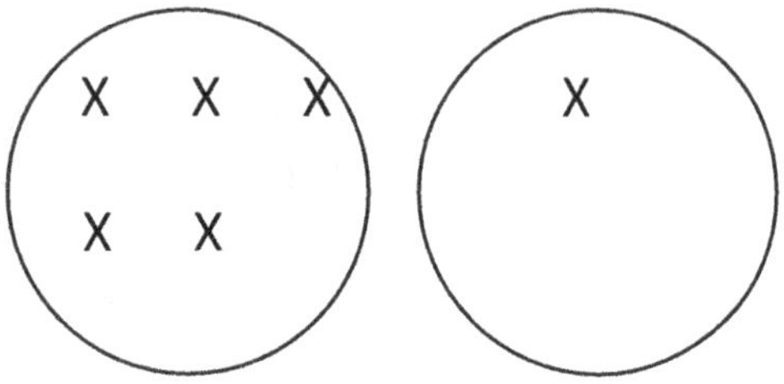

3 + 1

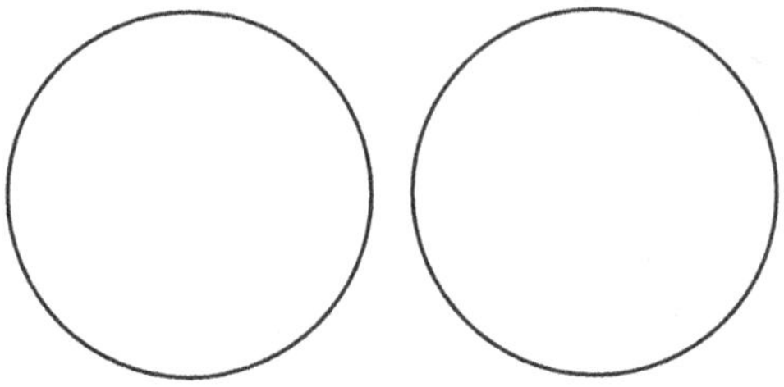

Write in a word sentence:

Example: 4 + 1 <u>four objects plus one object</u>

7 + 1 ________________________________

B

2 + 1 = ______	0 + 1 = ______	9 + 1 = ______	1 + 1 = ______	6 + 1 = ______
1 + 7 = ______	3 + 0 = ______	0 + 1 = ______	1 + 8 = ______	1 + 10 = ______
7 + 1 = ______	8 + 1 = ______	1 + 3 = ______	1 + 4 = ______	1 + 6 = ______
1 + 5 = ______	0 + 2 = ______	4 + 0 = ______	1 + 1 = ______	10 + 1= ______

Lesson C

C

Level 1

Draw a picture:

Example: 4 + 2

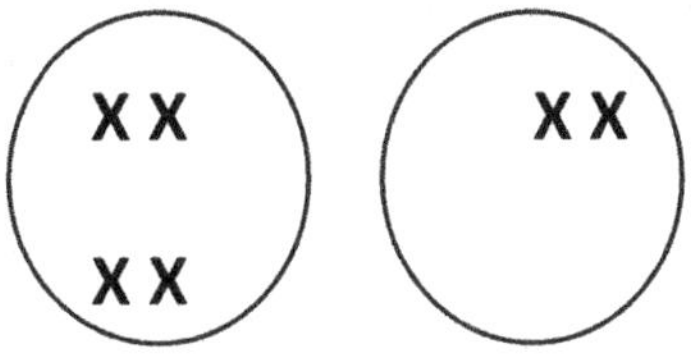

3 + 2

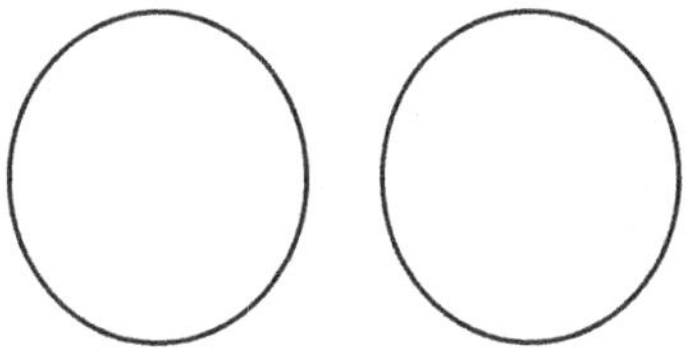

Write in a word sentence:

Example: 6 + 2 <u>six objects plus two objects</u>

10 + 2 ______________________________

New Facts: 2 + 2 2 + 4 4 + 2 **C**

7 + 1 = ______	0 + 8 = ______	1 + 3 = ______	2 + 4 = ______	0 + 6 = ______
2 + 2 = ______	1 + 5 = ______	4 + 2 = ______	1 + 1 = ______	6 + 1 = ______
1 + 4 = ______	3 + 0 = ______	0 + 1 = ______	1 + 8 = ______	2 + 2 = ______
2 + 4 = ______	1 + 9 = ______	4 + 2 = ______	2 + 1 = ______	10 + 1= ______

Lesson D

D

Level 1

Draw a picture:

4 + 2

3 + 2

Write in a word sentence:

Example: 3 + 2 <u>three objects plus two objects</u>

10 + 2 ______________________________

New Facts: 2 + 6 6 + 2 2 + 8 8 + 2

D

0 + 9 = ______	2 + 8 = ______	1 + 3 = ______	6 + 1 = ______	2 + 6 = ______
2 + 4 = ______	1 + 5 = ______	4 + 2 = ______	1 + 1 = ______	6 + 2 = ______
1 + 4 = ______	7 + 0 = ______	8 + 1 = ______	8 + 2 = ______	2 + 2 = ______
7 + 1 = ______	1 + 9 = ______	4 + 2 = ______	2 + 1 = ______	10 + 1= ______

Lesson E

Level 1

E

Draw a picture:

2 + 6

5 + 1

Write in a word sentence:

1 + 8 ________________________________

New Facts: 2 + 3   3 + 2   2 + 10   10 + 2

E

2 + 4 = ______	2 + 3 = ______	4 + 2 = ______	2 + 10 = _____	1 + 7= ______
0 + 5 = ______	2 + 8 = ______	0 + 3 = ______	2 + 2 = ______	2 + 6 = ______
2 + 1 = ______	10 + 2 = _____	9 + 1 = ______	3 + 2 = ______	6 + 2 = ______
1 + 4 = ______	6 + 1 = ______	8 + 1 = ______	8 + 2 = ______	2 + 2 = ______

Lesson F

Level 2

F

Draw a picture:

4 + 2

3 + 0

Write in a word sentence:

9 + 2 ______________________________

New Facts: 2 + 5   5 + 2   2 + 9   9 + 2

F

2 + 5 = ______	2 + 8 = ______	9 + 2 = ______	2 + 2 = ______	2 + 6 = ______
2 + 4 = ______	2 + 3 = ______	4 + 1 = ______	2 + 10 = _____	2 + 9= ______
2 + 4 = ______	10 + 2 = _____	4 + 2 = ______	2 + 3 = ______	6 + 2 = ______
5 + 2 = ______	6 + 1 = ______	8 + 1 = ______	8 + 2 = ______	0 + 2 = ______

Lesson G

G

Level 2

Draw a picture:

3 + 3

4 + 2

Write in a word sentence:

7 + 2 ________________________________

New Facts: 3 + 3 2 + 7 7 + 2

G

5 + 2 = ______	3 + 3 = ______	8 + 0 = ______	3 + 2 = ______	2 + 2 = ______
1 + 5 = ______	2 + 8 = ______	7 + 2 = ______	2 + 7 = ______	1 + 9 = ______
2 + 7 = ______	2 + 3 = ______	4 + 2 = ______	2 + 10 = _____	2 + 9= ______
2 + 4 = ______	7 + 2 = ______	1 + 2 = ______	2 + 6 = ______	6 + 2 = ______

Lesson H

Level 2

H

Write in a word sentence: 10 + 3

____

____

____

Draw a picture: 4 + 3

New Facts: 3 + 4 4 + 3 3 + 10 10 + 3

H

2 + 7 = ______	4 + 3 = ______	4 + 2 = ______	2 + 10 = ______	3 + 10 = ______
5 + 2 = ______	3 + 2 = ______	7 + 1 = ______	3 + 3 = ______	2 + 8 = ______
10 + 3 = ______	2 + 2 = ______	3 + 4 = ______	2 + 9 = ______	1 + 4 = ______
2 + 4 = ______	6 + 2 = ______	3 + 4 = ______	2 + 3 = ______	7 + 2 = ______

Lesson I

Level 2

I

Write in a word sentence: 4 + 8

____

____

____

Draw a picture: 2 + 5

New Facts: 3 + 6   6 + 3   3 + 8   8 + 3

I

3 + 3 = ______	3 + 6 = ______	1 + 4 = ______	2 + 9 = _____	2 + 2 = _____
3 + 6 = ______	3 + 4 = ______	2 + 5 = ______	10 + 2 = ____	6 + 3 = _____
6 + 2 = ______	3 + 2 = ______	3 + 8 = ______	3 + 6 = _____	3 + 8 = _____
8 + 3 = ______	7 + 2 = ______	1 + 9 = ______	4 + 3 = ______	3 + 10 = ____

Lesson J

Level 2

J

Write in a word sentence: 10 + 2

____

____

____

Draw a picture: 5 + 3

New Facts: 3 + 5   5 + 3   3 + 9   9 + 3

J

6 + 3 = ______	7 + 2 = ______	1 + 8 = ______	9 + 3 = ______	3 + 10 = ______
2 + 2 = ______	5 + 3 = ______	2 + 4 = ______	2 + 9 = ______	5 + 2 = ______
3 + 4 = ______	3 + 4 = ______	3 + 5 = ______	10 + 2 = ______	3 + 3 = ______
6 + 2 = ______	3 + 2 = ______	9 + 3 = ______	3 + 6 = ______	2 + 8 = ______

Lesson K

Level 3

K

Write in a word sentence: 2 + 7

____

____

____

Draw a picture: 3 + 7

New Facts: 3 + 7 7 + 3 4 + 4

K

6 + 0 = ______	3 + 7 = ______	9 + 3 = ______	3 + 8 = ______	2 + 8 = ______
4 + 4 = ______	7 + 2 = ______	4 + 4 = ______	3 + 3 = ______	3 + 6 = ______
7 + 3 = ______	5 + 3 = ______	2 + 4 = ______	2 + 9 = ______	5 + 2 = ______
4 + 4 = ______	3 + 4 = ______	3 + 10 = ______	10 + 2 = ______	7 + 3 = ______

Lesson L

Level 3

L

Write in a word sentence: 9 + 4

____

____

____

Draw a picture: 5 + 4

New Facts: 4 + 5   5 + 4   4 + 9   9 + 4

L

4 + 0 = ______	5 + 4 = ______	2 + 10 =______	9 + 4 = ______	7 + 3 = ______
5 + 2 = ______	3 + 3 = ______	9 + 3 =______	3 + 5 = ______	2 + 9 = ______
4 + 4 = ______	6 + 2 = ______	4 + 5 = ______	3 + 1 = ______	3 + 8 = ______
4 + 2 = ______	4 + 9 = ______	3 + 4 = ______	3 + 10 = ______	6 + 3 = ______

Lesson M

Level 3 **M**

Write in a word sentence: 9 + 3

____

____

____

Draw a picture: 6 + 4

Subtraction is the opposite of addition.

9 - 3 = ______ (Cross out 3 of the Os to make a subtraction picture.)

O O O O O O O O O

3 + _____ = 9

New Facts: 4 + 6   6 + 4   4 + 10   10 + 4 **M**

4 + 4 = ______	4 + 9 = ______	3 + 4 = ______	4 + 10 = _____	6 + 3 = _____
7 + 0 = ______	5 + 4 = ______	2 + 3 = ______	3 + 5 = ______	7 + 3 = _____
4 + 6 = ______	10 + 4 = _____	9 + 3 =______	3 + 8 = ______	2 + 2 = _____
3 + 10 = _____	8 + 2 = ______	1 + 5 = ______	6 + 4 = ______	2 + 7 = _____

Lesson N

Level 3 **N**

Write in a word sentence: 4 + 7

____

____

____

Draw a picture: 5 + 2

Subtraction is the opposite of addition.

11 - 4 = ______ (Cross out 4 of the Os to make a subtraction picture.)

O O O O O O O O O O O

4 + ______ = 11

New Facts: 4 + 7 7 + 4 4 + 8 8 + 4 **N**

3 + 6 = ______	8 + 3 = ______	7 + 1 = ______	7 + 4 = ______	6 + 4 = ______
4 + 7 = ______	4 + 9 = ______	4 + 4 = ______	2 + 10 = ____	6 + 2 = ______
7 + 3 = ______	8 + 4 = ______	3 + 3 = ______	3 + 5 = _____	3 + 4 = ______
4 + 5 = _____	10 + 4 = _____	9 + 3 = ______	4 + 8 = _____	0 + 9 = _____

Lesson O

Level 3 O

Write in a word sentence: 9 + 5

____

____

____

Draw a picture: 5 + 3

Subtraction is the opposite of addition.

10 - 2 = ______ (Cross out 2 of the Os to make a subtraction picture.)

O O O O O O O O O O

2 + _____ = 10

New Facts: 5 + 5 5 + 9 9 + 5 **O**

7 + 3 = ______	3 + 4 = ______	9 + 5 = ______	4 + 5 = _____	4 + 4 = ______
2 + 2 = _____	8 + 4 = _____	3 + 3 = ______	5 + 5 = _____	0 + 2 = ______
4 + 9 = ______	6 + 3 = ______	2 + 5 = ______	9 + 1 = ______	10 + 4 = _____
4 + 7 = ______	5 + 9 = ______	5 + 5 = ______	3 + 10 = ____	6 + 4 = _____

Lesson P

Level 4

P

Write in a word sentence: 5 + 10

____

____

____

Draw a picture: 2 + 5

Subtraction is the opposite of addition. Draw a subtraction picture.

5 - 3 = ______

O O O O O

3 + ____ = 5

New Facts: 5 + 6 6 + 5 5 + 8 8 + 5

P

2 + 8 = _____	8 + 5 = _____	9 + 4 = ______	5 + 5 = _____	0 + 8 = ______
3 + 9 = ______	6 + 5 = ______	2 + 7 = ______	8 + 3 = ______	8 + 4 = ______
4 + 7 = ______	5 + 9 = ______	5 + 8 = ______	4 + 10 = ____	6 + 4 = _____
4 + 1 = ______	5 + 4 = ______	5 + 3 = ______	5 + 6 = _____	4 + 4 = ______

Level 4

Q

Write in a word sentence: 9 + 4

____

____

____

Draw a picture: 3 + 4

Subtraction is the opposite of addition. Draw a subtraction picture.

8 - 4 = ______

O O O O O O O O

4 + ____ = 8

New Facts: 5 + 7 7 + 5 5 + 10 10 + 5 **Q**

5 + 0 = ______	5 + 7 = ______	8 + 3 = ______	5 + 5 = _____	3 + 3 = ______
2 + 9 = _____	8 + 5 = _____	9 + 4 =______	9 + 5 = _____	10 + 5 = _____
3 + 10 = _____	6 + 5 = ______	1 + 6 = ______	7 + 5 = ______	7 + 4 = ______
4 + 4 = ______	5 + 10 = _____	4 + 8 = ______	4 + 6 = ______	2 + 4 = _____

Level 4

R

Write in a word sentence: 6 + 9

____

____

____

Draw a picture: 2 + 5

Subtraction is the opposite of addition. Draw a subtraction picture.

6 - 2 = ______

O O O O O O

2 + ______ = 6

New Facts: 6 + 6   6 + 9   9 + 6

R

3 + 0 = ______	5 + 9 = ______	6 + 9 = ______	5 + 6 = ______	2 + 3 = ______
8 + 1 = ______	4 + 5 = ______	4 + 3 = ______	6 + 6 = ______	4 + 6 = ______
2 + 9 = ______	8 + 5 = ______	9 + 6 = ______	9 + 4 = ______	5 + 5 = ______
3 + 9 = ______	6 + 6 = ______	10 + 5 = ______	7 + 5 = ______	8 + 4 = ______

Lesson S

Level 4

S

Write in a word sentence: 5 + 8

____

____

____

Draw a picture: 4 + 3

Subtraction is the opposite of addition. Draw a subtraction picture.

7 - 4 = ______

O O O O O O O

4 + ____ = 7

New Facts: 6 + 7   7 +6   6 + 10   10 + 6

S

2 + 1 = _____	6 + 6 = ______	6 + 5 = ______	7 + 6 = ______	3 + 6 = _____
6 + 10 = ____	5 + 7 = _______	10 + 0 = _____	10 + 6 = _____	2 + 7 =_____
5 + 5 = ______	4 + 5 = ______	7 + 3 = ______	6 + 7 = _____	4 + 6 = _____
2 + 8 = _____	9 + 5 = _____	8 + 5 =______	4 + 4 = _____	3 + 5 = _____

Lesson T

Level 4 **T**

Write in a word sentence: 7 + 7

Draw a picture: 3 + 6

____

____

____

Subtraction is the opposite of addition. Make a subtraction picture.

10 - 5 = ______

O O O O O O O O O O

Write a subtraction equation to go with the diagram. _____ - _____ = _____

Complete the fact family for the numbers 5, 6, 11:

5 + 6 = _____

6 + 5 = _____

11 – 5 = _______

11 - 6 = _______

New Facts: 7 + 7 6 + 8 8 + 6 **T**

8 + 6 = _____	9 + 6 = _____	9 + 3 = ______	6 + 8 = _____	5 + 8 = ______
1 + 0 = _____	4 + 4 = ______	6 + 5 = ______	7 + 7 = ______	3 + 6 = ______
5 + 7 = ______	5 + 9 = ______	1 + 10 = ____-	10 + 6 = _____	2 + 4 = _____
6 + 6 = ______	4 + 7 = ______	7 + 7 = ______	6 + 7 = _____	9 + 4 = ______

Level 5

U

Write in a word sentence: 10 + 9

Draw a picture: 3 + 7

____

____

____

Subtraction is the opposite of addition. Make a subtraction picture.

12 - 3 = ______

O O O O O O O O O O O O

Write a subtraction equation to go with the diagram. _____ - _____ = _____

Complete the fact family for the numbers 3, 7, 10:

3 + 7 = ______

7 + ____ = 12

10 – 5 = ________

10 - 7 = _______

New Facts: 7 + 8   8 + 7   9 + 10   10 + 9

U

9 + 5 = ______	4 + 8 = ______	7 + 8 = ______	6 + 7 = _____	5 + 5 = ______
6 + 8 = _____	9 + 10 = _____	7 + 3 = ______	3 + 4 = _____	10 + 6 = _____
5 + 10 = ____	6 + 9 = ______	6 + 0 = ______	8 + 7 = ______	6 + 6 = ______
7 + 7 = ______	5 + 4 = _______	1 + 7 = ______	10 + 9 = _____	2 + 5 = _____

Lesson V

Level 5

V

Write in a word sentence: 9 + 6

____

____

____

Draw a picture: 4 + 5

Subtraction is the opposite of addition. Draw a subtraction picture.

14 - 5 = ______

O O O O O O O O O O O O O O

Write a subtraction equation to go with the diagram. _____ - _____ = _____

Complete the fact family for the numbers 4, 5, 9:

4 + _____ = 9

______ + 4 = 9

9 – 5 = ______

9 – 4 = ______

New Facts: 7 + 9   9 + 7   8 + 8

V

7 + 9 = ______	6 + 4 = ______	0 + 8 = ______	2 + 9 = ______	6 + 5 = ______
9 + 6 = ______	8 + 7 = ______	8 + 8 = ______	6 + 6 = ______	7 + 7 = ______
10 + 4 = ______	9 + 10 = ______	6 + 3 = ______	9 + 7 = ______	8 + 6 = ______
2 + 1 = ______	5 + 9 = ______	5 + 5 = ______	7 + 6 = ______	8 + 8 = ______

Level 5

W

Write in a word sentence: 6 + 5

____

____

____

Draw a picture: 2 + 8

Subtraction is the opposite of addition. Draw a subtraction picture.

8 - 3 = ______

O O O O O O O O

Write a subtraction equation to go with the diagram. _____ - _____ = _____

Complete the fact family for the numbers 2, 6, 8:

_____ + 6 = 8

6 + _____ = 8

8 – 2 = ______

_____ - 6 = 2

New Facts: 7 + 10 10 + 7 8 + 9 9 + 8

W

10 + 6 = ____	7 + 8 = ______	7 + 6 = ______	7 + 7 = ______	8 + 9 = ______
7 + 10 = _____	6 + 9 = ______	0 + 3 = ______	5 + 9 = ______	9 + 10 = _____
9 + 7 = ______	8 + 1 = ______	9 + 8 = ______	8 + 8 = _____	2 + 7 = ______
5 + 3 = _____	7 + 10 = _____	7 + 5 = ______	4 + 7 = _____	4 + 6 = ______

Lesson X

Level 5

X

Write in a word sentence: 6 + 3

Draw a picture: 4 + 7

____

____

____

Subtraction is the opposite of addition. Draw a subtraction picture.

6 - 4 = ______

O O O O O O

Write a subtraction equation to go with the diagram. _____ - _____ = _____

Complete the fact family for 3, 6, 9:

3 + _____ = 9

______ + 3 = 9

9 – 3 = ________

9 - ______ = 3

New Facts: 8 + 10 10 + 8 9 + 9 10 + 10

X

2 + 6 = _____	7 + 10 = _____	9 + 9 = ______	8 + 8 = _____	6 + 7 = ______
5 + 4 = _____	10 + 10 = _____	3 + 8 = ______	7 + 7 = ______	8 + 9 = ______
5 + 10 = _____	6 + 1 = _______	10 + 8 = _____	6 + 8 = ______	7 + 10 = _____
9 + 7 = ______	8 + 10 = ______	9 + 8 = ______	5 + 8 = _____	8 + 7 = ______

Lessson Y

Level 5 Y

Write in a word sentence: 6 + 7

____

____

____

Draw a picture: 4 + 3

Subtraction is the opposite of addition. Draw a subtraction picture.

12 - 5 = ______

O O O O O O O O O O O O

Write a subtraction equation to go with the diagram. _____ - _____ = ______

Complete the fact family for the numbers 5, 7, 12:

5 + _____ = 12

______ + 5 = 12

12 – 5 = ________

12 - ______ = 5

New Facts: None **Y**

7 + 7 = ______	8 + 9 = _______	3 + 5 = ______	4 + 8 = _____	3 + 7 = ______
2 + 3 = _____	7 + 10 = _____	9 + 9 =______	8 + 9 = _____	1 + 8 = ______
5 + 6 = _____	0 + 9 = ______	2 + 2 = ______	6 + 7 = ______	8 + 8 = ______
4 + 10 = _____	8 + 5 = _______	6 + 9 = _____	10 + 10 = _____	6 + 10 =_____

Lesson Z

Level 6

Z

Write in a word sentence: 4 + 8

____

____

____

Draw a picture: 2 + 6

Division is the opposite of multiplication. Draw a subtraction problem.

9 - 4 = ______

O O O O O O O O O

Write a subtraction equation to go with the diagram.

Complete the fact family for the numbers 4, 5, 9

4 + _____ = 9

5 + ______ = ______

9 – 4 = ________

9 - ______ = 4

New Facts: None

Z

10 + 10 = _____	8 + 6 = _______	6 + 6 = _____	4 + 4 = ______	3 + 8 = ______
7 + 7 = ______	8 + 8 = _______	9 + 8 = ______	9 + 8 = _____	3 + 6 = ______
2 + 6 = _____	0 + 10 = _____	9 + 9 =______	5 + 3 = _____	4 + 9 = ______
4 + 6 = _____	5 + 7 = ______	9 + 2 = ______	9 + 7 = ______	8 + 7 = ______

Mastery 1

10 + 10 = _____	8 + 6 = _______	6 + 6 = _____	4 + 4 = ______	3 + 8 = ______
7 + 7 = ______	8 + 8 = _______	9 + 8 = ______	0 + 8 = _____	3 + 6 = ______
2 + 6 = _____	0 + 10 = _____	9 + 9 =______	5 + 3 = _____	4 + 9 = ______
4 + 6 = _____	5 + 7 = ______	9 + 2 = ______	1 + 7 = ______	4 + 0 = ______
3 + 7 = _____	8 + 4 = ______	3 + 1 = ______	0 + 0 = _____	3 + 3 = ______
2 + 3 = _____	3 + 10 = _____	9 + 1 =______	8 + 9 = _____	1 + 8 = ______
5 + 6 = _____	0 + 9 = ______	2 + 2 = ______	6 + 7 = ______	8 + 8 = ______
4 + 10 = _____	8 + 5 = ______	6 + 9 = _____	10 + 5 = ______	6 + 10 =_____
2 + 6 = _____	3 + 10 = _____	5 + 9 =______	5 + 5 = _____	4 + 7 = ______
2 + 4 = _____	0 + 2 = ______	3 + 8 = ______	1 + 1 = ______	7 + 0 = ______
5 + 4 = _____	3 + 1 = ______	10 + 8 = _____	0 + 3 = ______	7 + 2 =_____
9 + 7 = _____	5 + 0 = ______	10 + 8 = _____	2 + 7 = _____	10 + 9 = _____
1 + 1 = _____	9 + 3 = ______	6 + 1 =______	5 + 4 = _____	4 + 3 = ____

Mastery 2

1 + 7 = ______	8 + 8 = ______	9 + 8 = ______	6 + 8 = ______	3 + 6 = ______
2 + 6 = ______	0 + 10 = ______	9 + 9 = ______	5 + 7 = ______	4 + 9 = ______
4 + 4 = ______	5 + 3 = ______	9 + 2 = ______	7 + 7 = ______	4 + 0 = ______
10 + 10 = ______	8 + 1= ______	6 + 6 = ______	4 + 8 = ______	3 + 8 = ______
3 + 7 = ______	6 + 4 = ______	3 + 1 = ______	0 + 0 = ______	3 + 10 = ______
2 + 3 = ______	3 + 3 = ______	9 + 1 = ______	5 + 9 = ______	6 + 8 = ______
5 + 5 = ______	0 + 9 = ______	2 + 2 = ______	6 + 7 = ______	8 + 8 = ______
4 + 10 = ______	4 + 5 = ______	6 + 9 = ______	10 + 5 = ______	6 + 10 = ______
2 + 6 = ______	3 + 10 = ______	8 + 9 = ______	5 + 6 = ______	0 + 7 = ______
2 + 4 = ______	0 + 2 = ______	3 + 8 = ______	3 + 1 = ______	7 + 2 = ______
5 + 8 = ______	1 + 1 = ______	6 + 8 = ______	0 + 3 = ______	7 + 4 = ______
9 + 7 = ______	5 + 0 = ______	10 + 8 = ______	2 + 7 = ______	10 + 9 = ______
1 + 1 = ______	9 + 3 = ______	6 + 1 = ______	5 + 4 = ______	4 + 3 = ______

Mastery 3

8 + 3 = _____	3 + 10 = _____	9 + 1 =______	6 + 9 = _____	1 + 8 = _____
5 + 6 = _____	0 + 9 = ______	2 + 2 = ______	6 + 2 = ______	8 + 8 = ____
4 + 10 = _____	5 + 5 = _______	8 + 9 = _____	10 + 5 = ______	6 + 10 =____
7 + 0 = _____	3 + 10 = _____	4 + 4 =______	8 + 5 = _____	4 + 7 = _____
2 + 4 = _____	0 + 2 = ______	3 + 2 = ______	8 + 8 = _____	7 + 6 = ______
10 + 10 = _____	1 + 7 = _______	6 + 6 = _____	7 + 5 = ______	3 + 8 = ____
7 + 7 = ______	1 + 1 = ______	7 + 5 = ______	0 + 4 = _____	3 + 6 = ______
2 + 6 = _____	0 + 10 = _____	0 + 0 =______	5 + 3 = _____	4 + 9 = _____
4 + 6 = _____	5 + 9 = ______	9 + 2 = ______	4 + 8 = ______	8 + 0 = _____
3 + 7 = ______	6 + 4 = _______	3 + 0 = ______	9 + 9 = _____	3 + 3 = _____
5 + 4 = ______	3 + 1 = _______	10 + 8 = _____	1 + 3 = ______	7 + 9 =_____
2 + 7 = ______	5 + 0 = ______	10 + 8 = _____	1 + 1 = _____	10 + 9 = ____
7 + 8 = _____	9 + 3 = ______	6 + 1 =______	5 + 4 = _____	4 + 3 = _____

Mastery 4

4 + 6 = _____	5 + 7 = _____	9 + 2 = _____	1 + 7 = _____	4 + 0 = _____
3 + 7 = _____	8 + 4 = _____	3 + 1 = _____	0 + 0 = _____	3 + 3 = _____
2 + 3 = _____	3 + 10 = _____	9 + 1 = _____	8 + 9 = _____	1 + 8 = _____
5 + 6 = _____	0 + 9 = _____	2 + 2 = _____	6 + 7 = _____	8 + 8 = _____
4 + 10 = _____	8 + 5 = _____	6 + 9 = _____	10 + 5 = _____	6 + 10 = _____
2 + 6 = _____	3 + 10 = _____	5 + 9 = _____	5 + 5 = _____	4 + 7 = _____
10 + 10 = _____	8 + 6 = _____	6 + 6 = _____	4 + 4 = _____	3 + 8 = _____
7 + 7 = _____	8 + 8 = _____	9 + 8 = _____	0 + 8 = _____	3 + 6 = _____
2 + 6 = _____	0 + 10 = _____	9 + 9 = _____	5 + 3 = _____	4 + 9 = _____
2 + 4 = _____	0 + 2 = _____	3 + 8 = _____	1 + 1 = _____	7 + 0 = _____
5 + 4 = _____	3 + 1 = _____	10 + 8 = _____	0 + 3 = _____	7 + 2 = _____
9 + 7 = _____	5 + 0 = _____	10 + 8 = _____	2 + 7 = _____	10 + 9 = _____
1 + 1 = _____	9 + 3 = _____	6 + 1 = _____	5 + 4 = _____	4 + 3 = _____

Mastery 5

2 + 4 = _____	0 + 2 = _____	3 + 8 = _____	1 + 1 = _____	7 + 0 = _____
5 + 4 = _____	3 + 1 = _____	10 + 8 = _____	0 + 3 = _____	7 + 2 = _____
9 + 7 = _____	5 + 0 = _____	10 + 8 = _____	2 + 7 = _____	10 + 9 = _____
1 + 1 = _____	9 + 3 = _____	6 + 1 = _____	5 + 4 = _____	4 + 3 = _____
7 + 10 = _____	8 + 6 = _____	6 + 6 = _____	4 + 4 = _____	3 + 8 = _____
7 + 7 = _____	8 + 8 = _____	9 + 8 = _____	0 + 8 = _____	3 + 6 = _____
2 + 6 = _____	0 + 10 = _____	9 + 9 = _____	5 + 3 = _____	4 + 9 = _____
4 + 6 = _____	5 + 7 = _____	9 + 2 = _____	1 + 7 = _____	4 + 0 = _____
3 + 7 = _____	8 + 4 = _____	3 + 1 = _____	0 + 0 = _____	3 + 3 = _____
2 + 3 = _____	3 + 10 = _____	9 + 1 = _____	8 + 9 = _____	1 + 8 = _____
5 + 6 = _____	0 + 9 = _____	2 + 2 = _____	6 + 7 = _____	8 + 8 = _____
4 + 10 = _____	8 + 5 = _____	6 + 9 = _____	10 + 5 = _____	6 + 10 = _____
2 + 6 = _____	3 + 10 = _____	5 + 9 = _____	5 + 5 = _____	4 + 7 = _____

There is no substitute for hard work.

Thomas Edison

New Facts Chart

Lessons A–M	Lessons N–Z
Lesson A: Numbers 1 to 10 plus 0	Lesson N: 4+7 7+4 4+8 8+2
Lesson B: Numbers 1 to 10 plus 1	Lesson O: 5+5 5+9 9+5
Lesson C: 2+2 2+4 4+2	Lesson P: 5+6 6+5 5+8 8+5
Lesson D: 2+6 6+2 2+8 8+2	Lesson Q: 5+7 7+5 5+10 10+5
Lesson E: 2+3 3+2 2+10 10+2	Lesson R: 6+6 6+9 9+6
Lesson F: 2+5 5+2 2+9 9+2	Lesson S: 6+7 7+6 6+10 10+6
Lesson G: 2+7 7+2 3+3	Lesson T: 7+7 6+8 8+6
Lesson H: 3+4 4+3 3+10 10+3	Lesson U: 7+8 8+7 9+10 10+9
Lesson I: 3+6 6+3 3+8 8+3	Lesson V: 7+9 9+7 8+8
Lesson J: 3+5 5+3 3+9 9+3	Lesson W: 7+10 10+7 8+9 9+8
Lesson K: 3+7 7+3 4+4	Lesson X: 8+10 10+8 9+9 10+10
Lesson L: 4+5 5+4 4+9 9+4	Lesson Y: None
Lesson M: 4+6 6+4 4+10 10+4	Lesson Z: None

New Facts Chart

(with answers)

Lesson A: Numbers 1 to 10 plus 0

Lesson B: Numbers 1 to 10 plus 1

Lesson C: 2+2=4   2+4=6   4+2=6

Lesson D: 2+6=8   6+2=8   2+8=10   8+2=10

Lesson E: 2+3=5   3+2=5   2+10=12   10+2=12

Lesson F: 2+5=7   5+2=7   2+9=11   9+2=11

Lesson G: 2+7=9   7+2=9   3+3=6

Lesson H: 3+4=7   4+3=7   3+10=13   10+3=13

Lesson I: 3+6=9   6+3=9   3+8=11   8+3=11

Lesson J: 3+5=8   5+3=8   3+9=12   9+3=12

Lesson K: 3+7=10   7+3=10   4+4=8

Lesson L: 4+5=9   5+4=9   4+9=13   9+4=13

Lesson M: 4+6=10   6+4=10   4+10=14   10+4=14

Lesson N: 4+7=11   7+4=11   4+8=12   8+2=12

Lesson O: 5+5=10   5+9=14   9+5=14

Lesson P: 5+6=11   6+5=11   5+8=13   8+5=13

Lesson Q: 5+7=12   7+5=12   5+10=15   10+5=15

Lesson R: 6+6=12   6+9=15   9+6=15

Lesson S: 6+7=13   7+6=13   6+10=16   10+6=16

Lesson T: 7+7=14   6+8=14   8+6=14

Lesson U: 7+8=15   8+7=15   9+10=19   10+9=19

Lesson V: 7+9=16   9+7=16   8+8=16

Lesson W: 7+10=17   10+7=17   8+9=17   9+8=17

Lesson X: 8+10=18   10+8=18   9+9=18   10+10=100

Lesson Y: None

Lesson Z: None

Vocabulary

Addition: finding the **total** by combining 2 or more numbers (groups)

Diagram: drawing that explains or shows something

Difference: answer to a subtraction problem

Fact Family: 2 addition and 2 subtraction problems using the same numbers

Equation: **number sentence** using numerals and symbols (Example: 4 + 2 = 6)

Sum: the **answer** to an addition problem

Total: the **answer** to an addition problem

Properties

Identity Property of Addition: adding zero to any number gives you the same number

Commutative Property: numbers can be added in any order and the answer will be the same

The difference between winning and losing is most often not quitting.

Walt Disney

Alternate Strategies

If you need to use alternate strategies, it is because your child is struggling. If that is the case, it is even more important to try to make it fun. People, especially children, put more effort into the things they enjoy. Avoid power struggles at all cost. No one wins.

1. Games are always a favorite. There are many computer games to practice facts. Just google addition facts. You can purchase games like Addition War or Bingo from Amazon or a teachers' supply stores.
2. Write silly sentences about a hard to remember fact and say them throughout the day. (Example: 8 + 7 = 15) Eight elephants carried 7 foxes to the zoo. Can you believe that is 15 new animals for the zoo!)
3. Draw a picture using the equation – put eyes on the 8, 7, and the 15. Repeat the fact. Make animals or cars or fruit, etc. out of the numbers. Then repeat the fact.
4. Flash Cards I (Make it fun. Put hard to remember facts on a card without the answer. Put a couple easy facts on the different cards.) Lay out cards one at a time. Have child say the problem and the answer. He gets the ones he knows. You get the ones he needs to practice. See who gets the most. Say each equation of all the ones you have. Play again with only the hard one to see how many more your child can get (add to group he or she already has). Make sure there are only 2 to 3 flashcards your child does not know. If there are more than that, return to the previous lesson or two. Presenting too many facts at once is one of the reason students confuse the answers.
5. Flash Cards II (Remember this is a game) Pull out all the facts your child knows from a set of flash cards. Add 2 or 3 facts your child is still learning. This game needs 3 people. Two to play game and one to hold up a card. The adult should have a 10 second handicap (wait time before answering). A card is held up. The first one to say the answer gets the card (don't forget wait time for adult – it is not fun if you lose all the time). Person with the most cards wins.

How to Use this Book in the Classroom

Key Concepts for Teachers:

- Fun, fun, fun – do all you can to keep this fun - use rewards, competitions, charts
- Students must pass one lesson before moving to the next lesson – high five, sticker, smiley face – acknowledge accomplishment – whatever works for your student. For the struggling child celebrate the growth (more correct answers on the page than yesterday or last week).
- Keeping track of progress – this is the biggest difference between home and school use. Obviously, this will be easier with a smaller class than a big class. Make a chart of all students using the program along with the Lessons. Fill in or check box for Lessons a student has completed with 100% correct.
- Material management – the easiest way I have found is to create a file box. This can be finished completely before you start the program, or you can add to file as needed. Label a manila folder with each Lesson – A, B, C, etc. For day 1, make enough copies of Lesson A for every student in your class, plus a couple extra. Do this for every lesson. If you are doing it as you go, when you make copies for Lesson B, make enough for everyone even if some students will not be starting Lesson B tomorrow. Also, make enough copies of Lesson A for students who need to repeat the Lesson. Be sure to always have enough copies for all students working on a particular level. When I was teaching, I wrote each student's name on the paper he or she was to do the next day.
- Teach new facts before students complete the student practice pages. Before the practice page is the time for students to figure out answers using various strategies to calculate the answers. Have students write problems in their assignment notebook. See Chart for new problems for each lesson. If you post this chart somewhere, it will make it easier. Another idea is to make a copy for each student and tape it into their assignment notebook.
- Student should know facts from memory when they do the student practice page (Lessons A – Z and Mastery 1 – 5)
- If student is on any level for more than 2 or 3 days, practice new facts in isolation (Note: it is more effective to practice for 2 minutes 5 times a day than it is to do 1 ten-minute practice session)
- Students can complete in writing or orally. In a classroom, you probably don't have to worry about students completing lesson orally unless you have a student with an IEP and oral accommodations.
- Teach vocabulary – another major problem for students is not understanding the vocabulary.

- One-size does not fit all, use additional strategies if have a student who is still struggling.

The top section of student practice page (Lessons A – Z) is understanding. The bottom section is memorization and should be timed. Timing is not critical, but it is used to keep students focused on the assignment. Depending on the age and manual dexterity of your students, anywhere from 30 – 60 seconds should be more than enough. Remember, it is not that the student can spit the answers back at lightning speed, but that the student is giving the answers from memory, not calculating the answers. Students can also answer orally. Although, one thing to note, research has shown that the physical act of writing triggers the memory area of the brain. Unless a student has a physical handicap, which makes writing difficult, writing will make the memorization process faster.

Key Ideas (teach these concepts as you work through the book) **for Student:**

- 3 + 5 is read as 3 plus 5
- 3 + 5 means 3 objects and 5 more objects
- Addition is reciprocal: 3 + 5 is the same answer as 5 + 3
- This is also the time to talk about adding things that are alike. You can add apples, or you can add oranges. If you need to add apples and oranges, you have to change the category to fruit. Talking about this now paves the way for future math concepts like adding fractions with unlike denominators.

Made in the USA
Monee, IL
08 November 2026

68002504R00033